STEM Superstars

Katherine Johnson

by Anita Nahta Amin

Norwood House Press

Cover: Katherine Johnson was a mathematician for NASA.

Norwood House Press

For information regarding Norwood House Press, please visit our website at: www.norwoodhousepress.com or call 866-565-2900.

PHOTO CREDITS: Cover: ©: Bob Nye/LRC/NASA; ©: Bob Nye/LRC/NASA, 9, 12; ©: CatLane/iStockphoto, 6; ©: Jordan Strauss/Invision/AP Images, 20; ©: Malachi Jacobs/Shutterstock Images, 5; ©: maroke/iStockphoto, 11; ©: NACA/LRC/NASA, 15; ©: NASA/Donaldson Collection/Michael Ochs Archives/Getty Images, 17; ©: NASA/Interim Archives/Archive Photos/Getty Images, 18

Hardcover ISBN: 978-1-68450-836-5
Paperback ISBN: 978-1-68404-632-4

Library of Congress Cataloging-in-Publication Data
Names: Amin, Anita Nahta, author.
Title: Katherine Johnson : NASA mathematician / by Anita Nahta Amin.
Description: Chicago, IL : Norwood House Press, [2021] | Series: STEM superstars | Includes index. | Audience: Grades K-1
Identifiers: LCCN 2019058380 (print) | LCCN 2019058381 (ebook) | ISBN 9781684508365 (hardcover) | ISBN 9781684046324 (paperback) | ISBN 9781684046362 (ebook)
Subjects: LCSH: Johnson, Katherine G.--Juvenile literature. | United States. National Aeronautics and Space Administration--Officials and employees--Juvenile literature. | African American women mathematicians--Biography--Juvenile literature. | African American teachers--Biography--Juvenile literature.
Classification: LCC QA29.J64 A45 2021 (print) | LCC QA29.J64 (ebook) | DDC 510.92 [B]--dc23
LC record available at https://lccn.loc.gov/2019058380
LC ebook record available at https://lccn.loc.gov/2019058381

328N—072020
Manufactured in the United States of America in North Mankato, Minnesota.

★ Table of Contents ★

Chapter 1

Early Life

Katherine Johnson was born in West Virginia in 1918. She had three older siblings.

Segregation made life hard. Black people were not treated well. But this did not stop her.

Katherine Johnson grew up in Greenbrier County in West Virginia.

Johnson was a fast learner. She loved to count.

Did You Know?

Johnson's dad was good at math. He knew she shared his talents.

Johnson loved to learn. She asked questions. She made up counting games. She started high school at age ten.

Johnson studied French and math in college. She worked hard. Teachers made classes just for her. One class was the **geometry** of space. She studied **orbit** shapes.

Johnson used math throughout her life.

Chapter 2

First Jobs

Johnson took a job as a teacher. She taught French, math, and music. She wanted kids to like learning.

Johnson wanted to be a **mathematician**. But this job was rare for black women.

Johnson worked as a teacher after college.

A friend told Johnson about the **National Advisory Committee for Aeronautics (NACA)**. NACA studied aircraft. Johnson could find a math job there.

But the jobs were full. Johnson tried for one anyway. NACA hired her. She became a mathematician.

Johnson wanted to be a mathematician. She never gave up on her dream.

Did You Know?

NACA math workers were called computers.

Johnson tested new kinds of planes. She measured **data** like air flow. It showed whether a plane was safe. She also studied plane crashes.

Johnson used math to test planes for NACA.

Chapter 3

Space Hero

The country wanted to send people into space. NACA held meetings. Only men were allowed. But this did not stop Johnson. She asked to join them.

Johnson worked hard at her job.

Johnson (right) became very important to others working at NASA.

NACA became the **National Aeronautics and Space Administration (NASA)**. Workers built a rocket. Johnson figured out its flight path. Then the first American flew to space.

Did You Know?
Johnson had three daughters.

Later, machines did the math. John Glenn had doubts. He was going to space. He asked for Johnson. He trusted her math. She mapped his flight. Glenn became the first American to orbit Earth.

Then Johnson helped the first man reach the moon. Later she helped save the Apollo 13 crew. Their rocket caught fire. Her flight path took them home.

She won medals. Two NASA buildings are named after her. There are also books and a movie about her.

Johnson blazed a path for all. Nothing stopped this math star from achieving her dreams.

In 2016, Johnson attended an event celebrating the movie *Hidden Figures*. The movie told Johnson's story.

★ Career Connections ★

1 Johnson loved to count. Count things you see around you. Count by ones, fives, tens, or any number. You can even keep a list of what you have counted.

2 Learn math. Make up your own math problems to solve. Or ask an adult to make math problems for you. Read math books. Join a math club or start your own.

3 Learn about space. Read books about space. Visit a planetarium. Buy or borrow a telescope to look at the moon, stars, and planets. There are even space camps for kids.

4 Don't be scared to ask questions. If you want to know how something works, ask someone. Johnson asked many questions. This is how she learned so much.

Glossary

data (DAY-tuh): Information, like measurements, collected for study.

geometry (jee-AH-meh-tree): The study in math of shapes, lines, and angles.

mathematician (MATH-uh-meh-tih-shun): A person who uses math to do their job.

National Advisory Committee for Aeronautics (NACA) (NAK-uh): The United States government group that was in charge of flight research.

National Aeronautics and Space Administration (NASA) (NASS-uh): The United States government group in charge of exploring space.

orbit (OR-bit): The path around an object, such as a planet or star, in space.

segregation (seg-ruh-GAY-shun): The separation of people based on the color of their skin.

For More Information

Books

Joe Levit. *Let's Explore Math*. Minneapolis, MN: Lerner Publications, 2018. This book teaches math concepts.

K. C. Kelley. *Space*. Mankato, MN: Amicus, 2018. This book gives an overview of outer space.

Websites

Khan Academy: Early Math

(https://www.khanacademy.org/math/early-math) This website teaches many basic math concepts.

NASA Kids' Club

(https://www.nasa.gov/kidsclub/index.html) This website teaches kids about space and NASA.

★ Index ★

★ About the Author ★

Anita Nahta Amin is an Indian American author of children's fiction and nonfiction books. Many of her stories also appear in children's magazines and educational resources. She has degrees in biomedical engineering, electrical engineering, and computer engineering. A former information technology manager and business consultant, she now enjoys writing about STEM topics for children. She lives in Florida with her husband and twin children. She enjoys using math to write secret coded messages for her kids.